Editor
Mary S. Jones, M.A.

Cover Artist
Delia Rubio

Editor in Chief
Karen J. Goldfluss, M.S. Ed.

Illustrator
Greg Anderson-Clift

Art Production Manager
Kevin Barnes

Imaging
Leonard P. Swierski

Publisher

Mary D. Smith, M.S. Ed.

Author

Bev Dunbar

Teacher Created Resources, Inc.
6421 Industry Way
Westminster, CA 92683
www.teachercreated.com

ISBN: 978-1-4206-8105-5

© *2008 Teacher Created Resources, Inc.*
Made in U.S.A.

Table of Contents

Introduction

The delightful illustrations and short, simple exercises in the *Early Math Skills* series will help young learners develop essential math skills with confidence. Each standards-based activity focuses on a specific skill. Clear instructions and examples will guide teachers and parents to help children complete the lessons successfully. Since each page includes a suggestion for extending the learning and reinforcing the skill, the books are ideal for any setting—a classroom, small-group tutoring, or at-home learning.

What's in This Book?

After students have learned to count up to 10 and can recognize and write numbers from 1 to 10, they can take the next step—adding and subtracting. Through using the activities in this book, students will:

- review counting and writing numbers from 1 to 10
- identify and draw patterns and groups of particular numbers
- add and subtract objects up to a total of 10

Work actively with students through each activity so that they understand what they are expected to do on each page. Read the instructions for each activity aloud to the students and model an example. Use the certificates on the next page to reward students for their hard work after they have completed a majority of the activities.

Features of Pages

INSTRUCTIONS — What students need to do for the activity.

EXAMPLE — The first one is done for students so they can see exactly what to do.

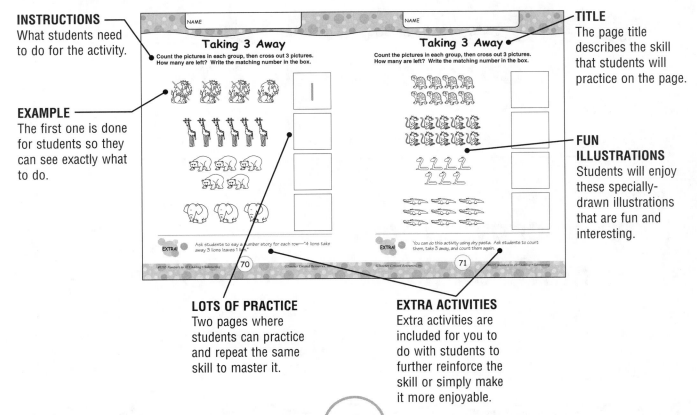

TITLE — The page title describes the skill that students will practice on the page.

FUN ILLUSTRATIONS — Students will enjoy these specially-drawn illustrations that are fun and interesting.

LOTS OF PRACTICE — Two pages where students can practice and repeat the same skill to master it.

EXTRA ACTIVITIES — Extra activities are included for you to do with students to further reinforce the skill or simply make it more enjoyable.

This award goes to

_____

for working
so hard!

Date _____

Congratulations to

_____

for improving your
math skills!

Date _____

Standards and Benchmarks

The activities in this book meet the following standards, which are used with permission from McREL.

Copyright 2006 McREL. Mid-continent Research for Education and Learning.
Address: 2250 S. Parker Road, Suite 500, Aurora, CO 80014
Telephone: 303-377-0990 Website: *www.mcrel.org/standards-benchmarks*

Standard 2. Understands and applies basic and advanced properties of the concepts of numbers

Level Pre-K (Grades Pre-K)
1. Understands that numbers represent the quantity of objects
2. Counts by ones to ten or higher
3. Counts objects
4. Understands one-to-one correspondence
5. Understands the concept of position in a sequence (e.g., first, last)
6. Knows the written numerals 0-9
7. Knows the common language for comparing quantity of objects (e.g., "more than," "less than," "same as")
8. Understands that a whole object can be separated into parts
9. Knows that numbers are used in real-world situations

Level I (Grades K-2)
1. Understands that numerals are symbols used to represent quantities or attributes of real-world objects
2. Counts whole numbers (i.e., both cardinal and ordinal numbers)
3. Understands symbolic, concrete, and pictorial representations of numbers (e.g., written numerals, objects in sets, number lines)
4. Understands basic whole number relationships (e.g., 4 is less than 10, 30 is 3 tens)

Standard 3. Uses basic and advanced procedures while performing the processes of computation

Level Pre-K (Grades Pre-K)
1. Knows that the quantity of objects can change by adding or taking away objects

Level I (Grades K-2)
1. Adds and subtracts whole numbers
2. Solves real-world problems involving addition and subtraction of whole numbers

Standard 8. Understands and applies basic and advanced properties of functions and algebra

Level Pre-K (Grades Pre-K)
1. Understands simple patterns (e.g., boy-girl-boy-girl)
2. Repeats simple patterns

Reviewing Numbers 1-5

Look at the cards hanging on the lines. How many pictures are on the first card? Write the matching number on the middle card, and draw the same number of triangles in the last card.

 **EXTRA!** Ask students to name the shapes they are drawing in the bottom row.

Reviewing Numbers 1-5

Look at the cards hanging on the lines. How many pictures are on the first card? Write the matching number on the middle card, and draw the same number of triangles in the last card.

 **EXTRA!** *Point to a number. Ask students to collect that many objects from the room.*

Making Patterns of 6

Pick two different colors. Color the first 6 shapes in one color, then the next 6 shapes in the other color.

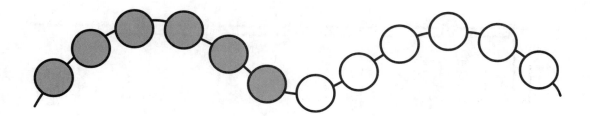

 **EXTRA!** Ask students to count the fish on this page.

Making Patterns of 6

Pick two different colors. Color the first 6 shapes in one color, then the next 6 shapes in the other color.

EXTRA!  You can help students make repeating patterns of 6 with objects around the classroom or with movements such as 6 jumps, 6 steps, 6 jumps, and so on.

9

Drawing Groups of 6

Draw matching pictures or shapes so there are 6 things in each group.

6 balloons

6 legs

 **EXTRA!** See if students can see any other groups of 6 around you in the classroom. Are there 6 chairs that look the same? Or 6 books or papers?

#8105 Numbers to 10 • Adding • Subtracting ©*Teacher Created Resources, Inc.*

Drawing Groups of 6

Draw matching pictures or shapes so there are 6 things in each group.

6 spots

6 triangles

 **EXTRA!**   Ask students to hold up 6 fingers and count them.

#8105 Numbers to 10 • Adding • Subtracting

Writing the Number 6

How do you draw the number 6? Start at the top and drag your pencil down to the left and around. Trace over all the number 6s below.

 **EXTRA!** Students can practice drawing large number 6s in the air before writing them in.

Writing the Number 6

How do you draw the number 6? Start at the top and drag your pencil down to the left and around. Trace over all the number 6s below. Then draw in six snakes.

6 snakes

 **EXTRA!** See how many number 6s students can find around the classroom.

©Teacher Created Resources, Inc.   #8105 Numbers to 10 • Adding • Subtracting

Making Patterns of 7

Pick two different colors. Color the first 7 shapes in one color, then the next 7 shapes in the other color.

EXTRA!   Ask students to hold up 7 fingers and count them.

Making Patterns of 7

Pick two different colors. Color the first 7 shapes in one color, then the next 7 shapes in the other color.

EXTRA!  Ask students to point to the group which has the most pictures in it. Check by counting.

15

Drawing Groups of 7

Draw matching pictures so there are 7 things in each group.

7 kittens

7 fish

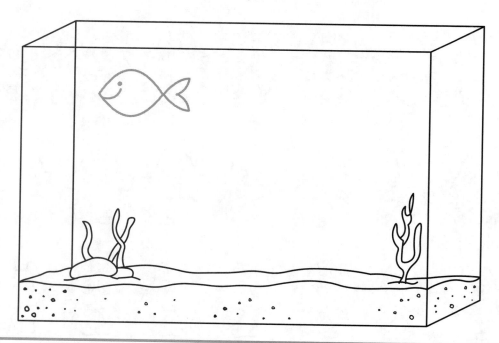

EXTRA!

Encourage students to count whenever they can.
How many steps can they take across a room?
How many spoons are laid out on the table?
How many birds can they see outside?

Drawing Groups of 7

Draw matching pictures so there are 7 things in each group.

7 glasses

7 leaves

 **EXTRA!**   Are there 7 glasses and 7 leaves? Check by counting.

Writing the Number 7

How do you draw the number 7? Start at the top left. Drag your pencil across to the right and down to the left. Trace over all the number 7s below.

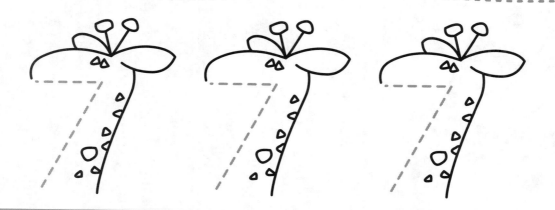

 **EXTRA!**

Students can practice drawing large number 7s in the air, before writing them in.

18

Writing the Number 7

How do you draw the number 7? Start at the top left. Drag your pencil across to the right and down to the left. Trace over all the number 7s below. Then draw in 7 sheep.

7 sheep

 **EXTRA!**

Ask students to stamp their feet 7 times and clap their hands 7 times. What else can they do 7 times?

Reviewing Numbers 1-7

Look at the cards hanging on the lines. How many pictures are on the first card? Write the matching number on the middle card, and draw the same number of circles in the last card.

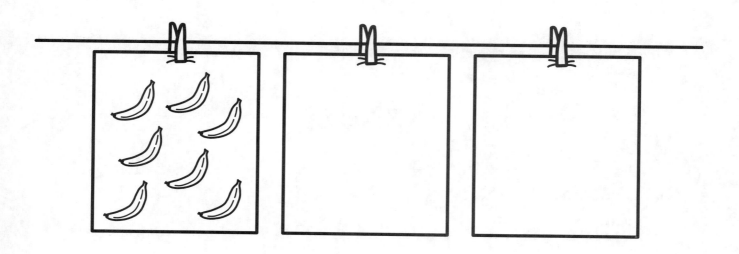

Find 6 forks and match them to 6 spoons.

Reviewing Numbers 1-7

Look at the cards hanging on the lines. How many pictures are on the first card? Write the matching number on the middle card, and draw the same number of circles in the last card.

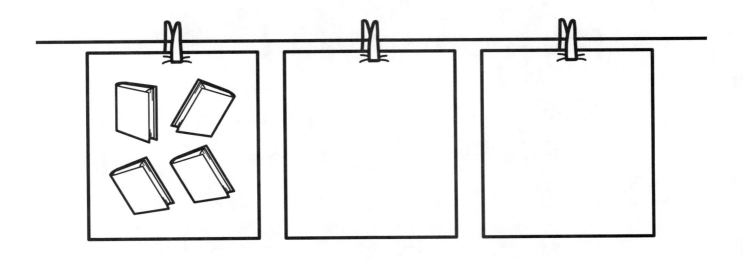

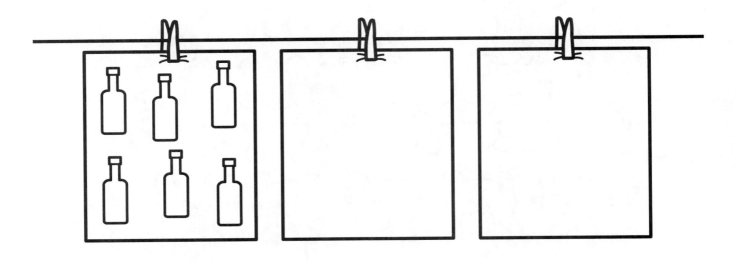

EXTRA!  Find 7 pencils and match them to 7 crayons.

#8105 Numbers to 10 • Adding • Subtracting

Making Patterns of 8

Pick two different colors. Color the first 8 shapes in one color, then the next 8 shapes in the other color.

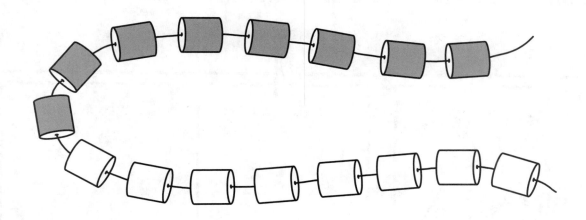

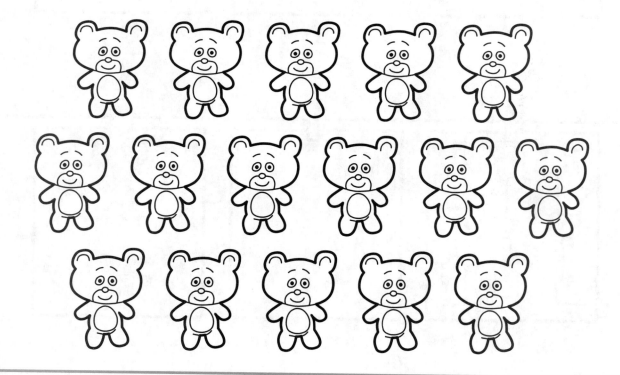

EXTRA!   Ask students to hold up 8 fingers and count them.

Making Patterns of 8

Pick two different colors. Color the first 8 shapes in one color, then the next 8 shapes in the other color.

  Ask students to nod their heads 8 times, and shake their heads 8 times. What else can they do 8 times?

#8105 Numbers to 10 • Adding • Subtracting

Drawing Groups of 8

Draw matching pictures so there are 8 things in each group.

8 hairs

8 bananas

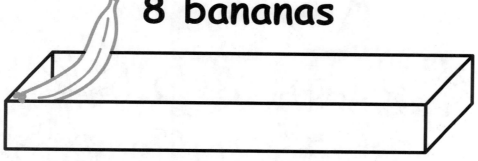

8 wheels

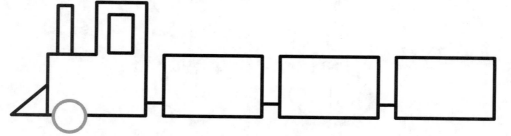

 **EXTRA!** Help students count to 8 by counting along with a simple song or piece of music.

Drawing Groups of 8

Draw matching pictures so there are 8 things in each group.

8 carrots

8 spots

8 squares

Ask students to point to their 8th finger.

Writing the Number 8

How do you draw the number 8? Start at the top right. Curve your pencil around to the left and down to the bottom right in a smooth curve, then back up to join where you started. Trace over all the number 8s below.

 **EXTRA!** Students can practice drawing large number 8s in the air before writing them in.

Writing the Number 8

How do you draw the number 8? Start at the top right. Curve your pencil around to the left and down to the bottom right in a smooth curve, then back up to join where you started. Trace over all the number 8s below. Then draw in 8 suns.

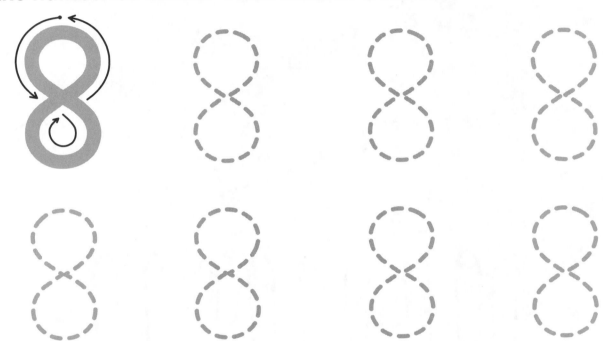

8 suns

 **EXTRA!** Draw 8 large squares on the board. Ask students to draw 8 sticks inside each square.

#8105 Numbers to 10 • Adding • Subtracting

Making Patterns of 9

Pick two different colors. Color the first 9 shapes in one color, then the next 9 shapes in the other color.

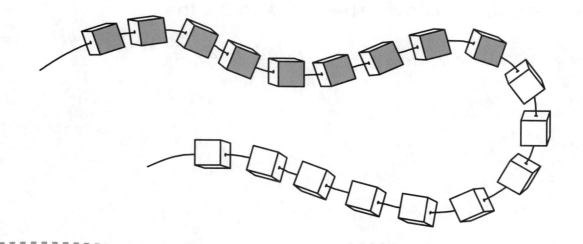

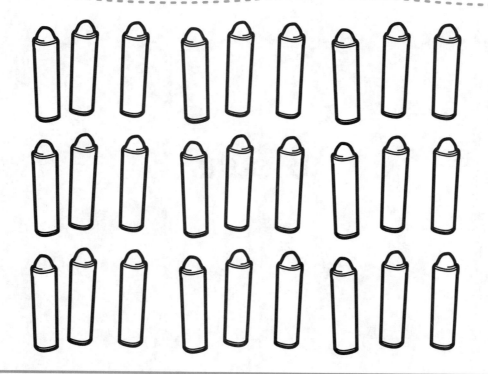

 **EXTRA!**   Ask students to hold up 9 fingers and count them.

Making Patterns of 9

Pick two different colors. Color the first 9 shapes in one color, then the next 9 shapes in the other color.

EXTRA! *Can students count how many sailboats are on this page?*

©*Teacher Created Resources, Inc.*   #8105 Numbers to 10 • Adding • Subtracting

Drawing Groups of 9

Draw matching pictures so there are 9 things in each group.

9 olives

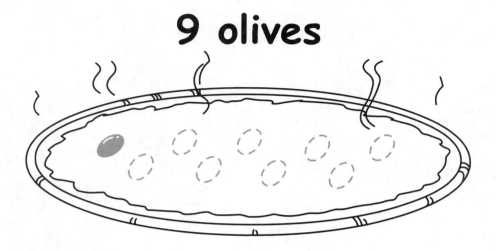

9 stripes

EXTRA!

Ask students to guess how far away they would be if they walked 9 steps. Then they can check.

30

Drawing Groups of 9

Draw matching pictures so there are 9 things in each group.

9 eggs

9 books

EXTRA! Ask students to guess how long 9 books placed end-to-end would be. Then they can check.

#8105 Numbers to 10 • Adding • Subtracting

Writing the Number 9

How do you draw the number 9? Start at the top right and drag your pencil around to the left to make a circle. Then drag your pencil down to make a stick. Trace over all the number 9s below.

EXTRA!

Students can practice drawing large number 9s in the air before writing them in.

Writing the Number 9

How do you draw the number 9? Start at the top right and drag your pencil around to the left to make a circle. Then drag your pencil down to make a stick. Trace over all the number 9s below. Then draw in 9 nuts.

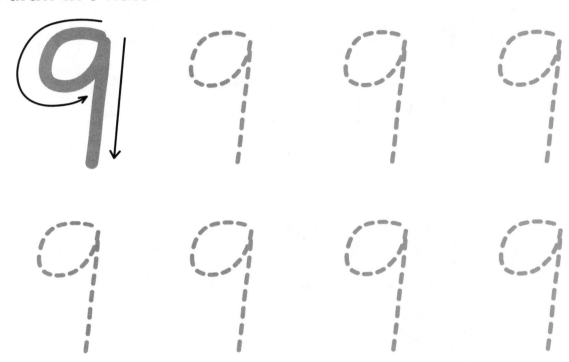

9 nuts

EXTRA! Ask students if they can stand on one leg for 9 seconds and have them count with you.

33

Making Patterns of 10

Pick two different colors. Color the first 10 shapes in one color, then the next 10 shapes in the other color.

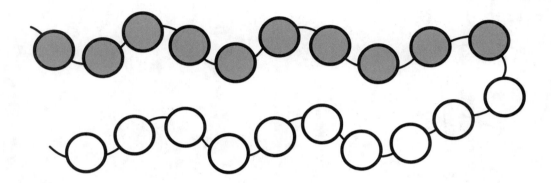

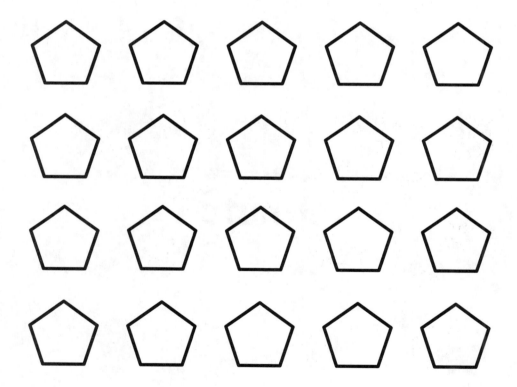

 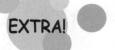 Ask students to hold up 10 fingers and count them.

Making Patterns of 10

Pick two different colors. Color the first 10 shapes in one color, then the next 10 shapes in the other color.

 **EXTRA!**   Sing songs or rhymes which involve the number 10 to help students count.

Drawing Groups of 10

Draw matching pictures so there are 10 things in each group.

10 fingers

10 circles

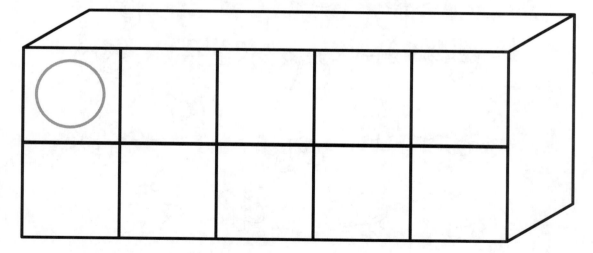

 Ask students to collect 10 pencils and match them to 10 pieces of paper.

Drawing Groups of 10

Draw matching pictures so there are 10 things in each group.

10 teeth

10 fries

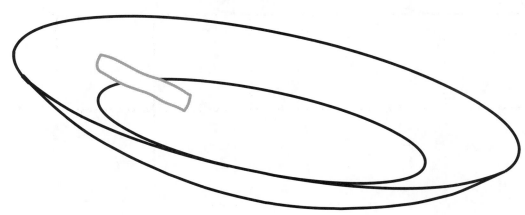

EXTRA! Ask students if there are 10 fries on the plate? 10 teeth? Count all the things on the page.

#8105 Numbers to 10 • Adding • Subtracting

NAME

Writing the Number 10

How do you draw the number 10? First draw a 1, starting at the top and going down in a straight line. To draw a 0, start at the top right and drag your pencil around to the left to make a circle. Trace over all the number 10s below.

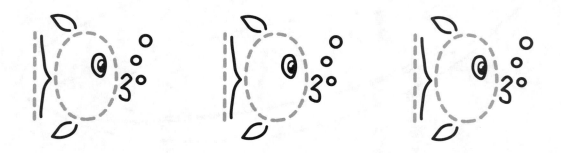

 **EXTRA!**   Students can practice drawing large number 10s in the air before writing them in.

38

#8105 Numbers to 10 • Adding • Subtracting   ©Teacher Created Resources, Inc.

Writing the Number 10

How do you draw the number 10? First draw a 1, starting at the top and going down in a straight line. To draw a 0, start at the top right and drag your pencil around to the left to make a circle. Trace over all the number 10s below. Then draw in 10 tops.

10 spinning tops

Ask students to see if they can count to 10 in less than 10 seconds.

©Teacher Created Resources, Inc.   #8105 Numbers to 10 • Adding • Subtracting

Joining the Matching Sets

How many animals are in each box? Draw a line between the boxes that have the same number of animals inside them.

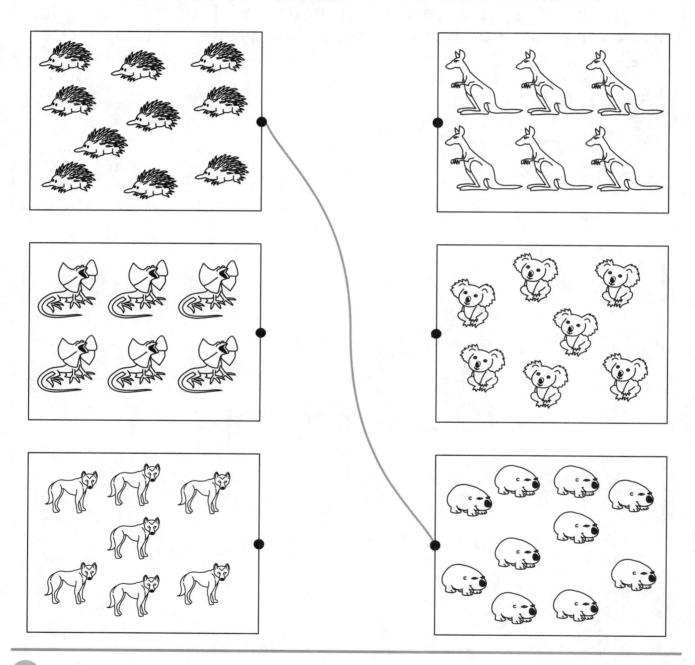

 **EXTRA!**

Ask students if there are more kangaroos than koalas.
Then count them to check.

#8105 Numbers to 10 • Adding • Subtracting

Joining the Matching Sets

How many animals are in each box? Draw a line between the boxes that have the same number of animals inside them.

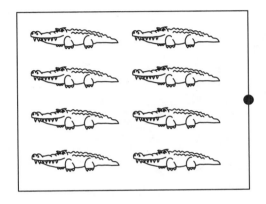

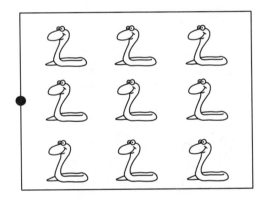

 **EXTRA!**

Ask students if there are more tigers than snakes.
Then count them to check.

41

Finding the Matching Number

How many pictures can you see in each box? Circle the matching number.

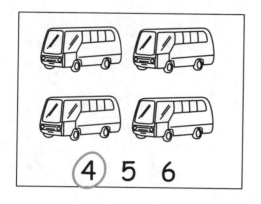

4 5 6

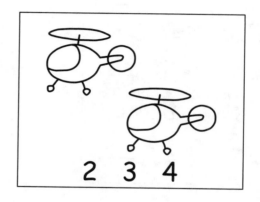

2 3 4

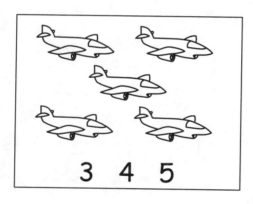

3 4 5

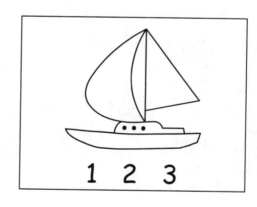

1 2 3

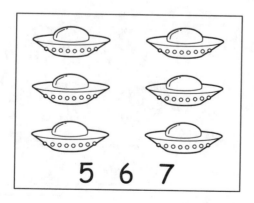

5 6 7

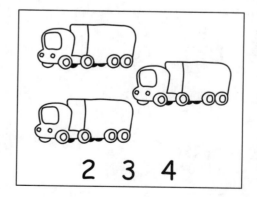

2 3 4

EXTRA! Ask students if there are more jets or more spaceships. Then count them to check.

42

©*Teacher Created Resources, Inc.*

Finding the Matching Number

How many pictures can you see in each box? Circle the matching number.

8   9   10

6   7   8

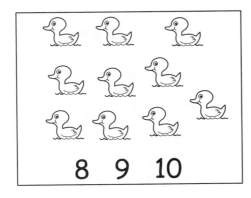

8   9   10

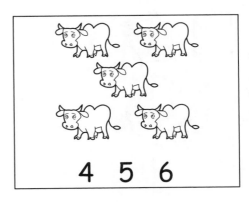

4   5   6

7   8   9

5   6   7

EXTRA!

Ask students if there are more ducks or more sheep.
Then count them to check.

#8105 Numbers to 10 • Adding • Subtracting

Writing the Matching Number

Count the pictures in each group, then write the matching number beside them.

2

_____

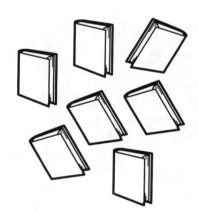

_____

_____

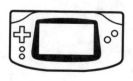

_____

_____

 **EXTRA!** You can play counting games with real objects—ask students to count a group of toys, or a pile of books, with up to 10 in the pile.

Writing the Matching Number

Count the pictures in each group, then write the matching number beside them.

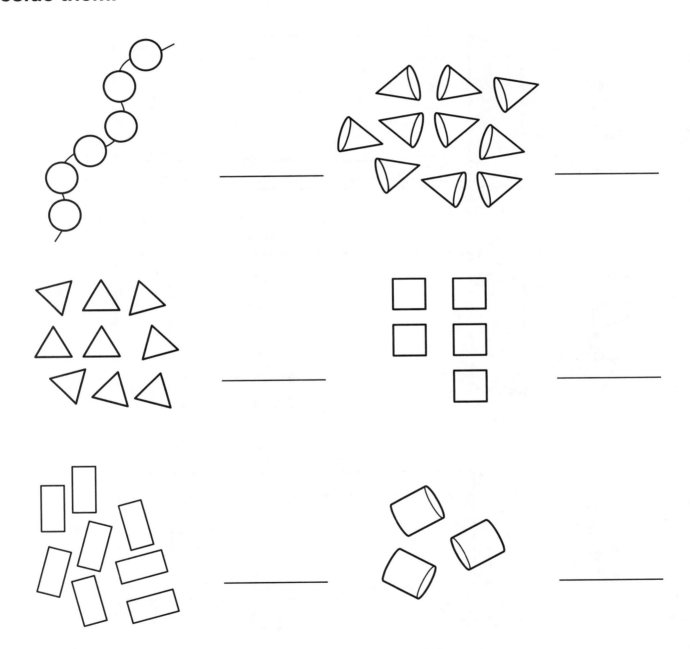

_____   _____

_____   _____

_____   _____

 **EXTRA!**   Ask students to write five number 5s, seven number 7s and three number 3s.

45

Drawing the Matching Sets

Draw fish to match the number beside each tank.

5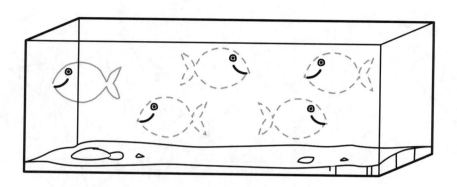

8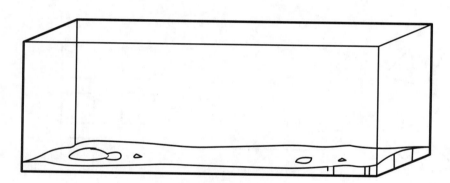

6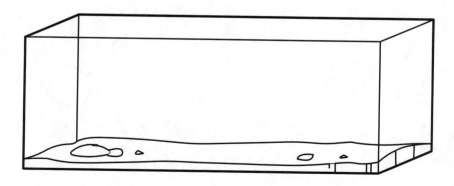

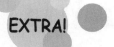 **EXTRA!** Ask students which tank has the smallest number of fish.

Drawing the Matching Sets

Draw spots on each cow to match the number beside each one.

7

10

9

 **EXTRA!**   Ask students which cow has the smallest number of spots.

47

Joining the Dots

Here is a hidden picture. Join the set of numbers from 1 to 10 to find the hidden picture. Be sure to start at number 1.

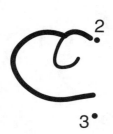
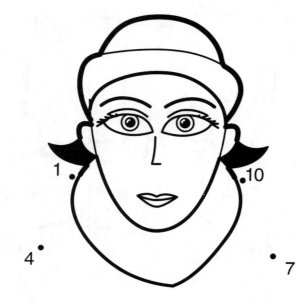
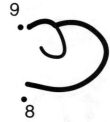
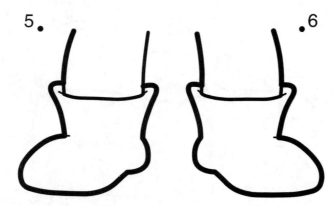

EXTRA! Put the numbers 1 to 10 on separate cards, shuffle them, and ask students to put them in the correct order.

Joining the Dots

Here is a hidden picture. Join the set of numbers from 1 to 10 to find the hidden picture. Be sure to start at number 1.

EXTRA!

Where else can students see the numbers 1 to 10 in order? On the telephone?

#8105 Numbers to 10 • Adding • Subtracting

Coloring by Numbers

Pick 10 different colored pencils or crayons. Color in the circles at the top of the page, then color the pictures below to match.

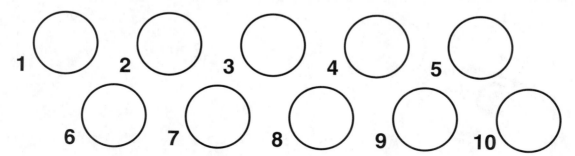

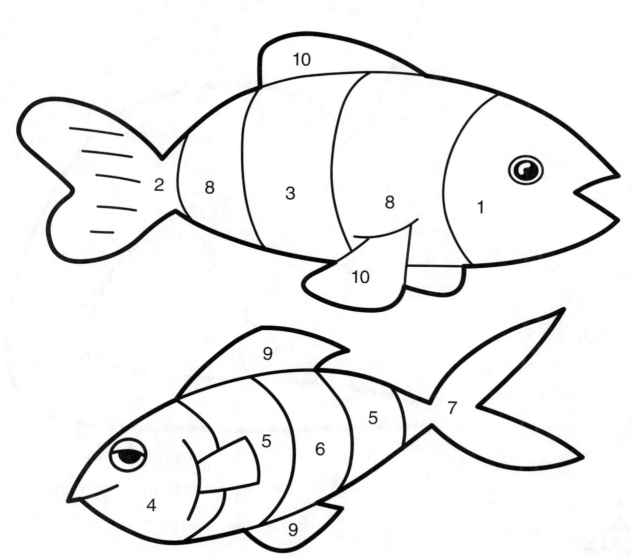

Coloring by Numbers

Pick 10 different colored pencils or crayons. Color in the circles at the top of the page, then color the picture below to match.

51

Counting and Writing 1–10

Count the baby animals in each basket. Write the number of animals in the box.

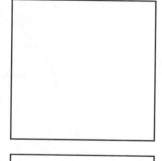

EXTRA!

Before you start, go over counting to 10 with students. Can they count to 10 on their fingers? To be able to add and subtract numbers, they will need to be confident in their counting.

Counting and Writing 1–10

Count the baby animals in each basket. Write the number of animals in the box.

EXTRA! Ask students to count all the rabbits on both pages. How many are there? How many puppies?

©*Teacher Created Resources, Inc.* #8105 Numbers to 10 • Adding • Subtracting

Drawing 1-10 Objects

Look at the number on each apple tree. Draw the matching number of apples.

EXTRA! Bring in a fruit bowl full of fruit. Ask students to count each different type of fruit in it.

Drawing 1-10 Objects

Look at the number on each apple tree. Draw the matching number of apples.

 **EXTRA!**

Ask students to count the apples on one tree and collect the matching number of objects from the room.

#8105 Numbers to 10 • Adding • Subtracting

Adding 1 More

Draw 1 more cookie in each jar. Count up how many cookies there are now and write this number next to the jar.

 **EXTRA!**

Help students to say a number story for each jar—"3 cookies and 1 more makes 4."

Adding 1 More

Draw 1 more cookie in each jar. Count up how many cookies there are now and write this number next to the jar.

 **EXTRA!**   Ask students which jar has the most cookies now. Color it in.

(57)

Adding 2 More

How many people are in each group? Draw 2 more in each group, count how many there are altogether, then write the number on the card.

 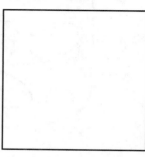

 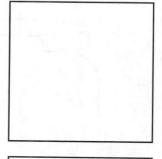

 **EXTRA!** Ask students to say a number story for each row—"2 babies and 2 more makes 4."

Adding 2 More

How many people are in each group? Draw 2 more in each group, count how many there are altogether, then write the number on the card.

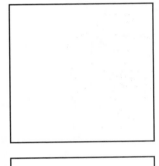

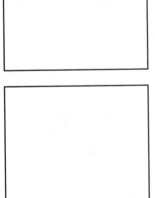

The last row on this page has no pictures. Help students to draw 2 faces, count them, and write the number in the box.

#8105 Numbers to 10 • Adding • Subtracting

Adding 3 More

How many pictures are in each group? Draw 3 more in each group, count how many there are altogether, then write the number in the box.

 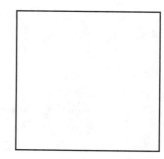

 **EXTRA!** Ask students to say a number story for each row—"2 spoons and 3 more make 5."

Adding 3 More

How many pictures are in each group? Draw 3 more in each group, count how many there are altogether, then write the number in the box.

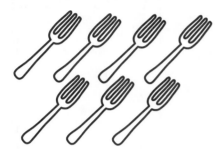

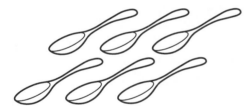

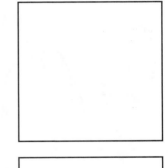

 **EXTRA!** The last row on this page has no pictures. Help students to draw 3 plates, count them, and write the number in the box.

#8105 Numbers to 10 • Adding • Subtracting

Adding 4 More

How many pictures are in each group? Draw 4 more in each group, count how many there are altogether, then write the number in the box.

 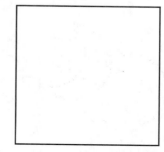

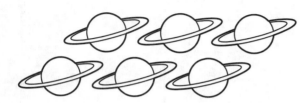 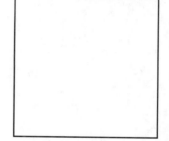

 **EXTRA!**

Don't forget to help students make a number story for each group. This becomes especially important as the numbers get larger.

Adding 4 More

How many pictures are in each group? Draw 4 more in each group, count how many there are altogether, then write the number in the box.

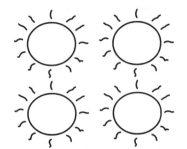

EXTRA! The last row on this page has no pictures. Help students draw 4 planets, then count them and write the number in the box.

63

NAME

Adding 5 More

How many pictures are in each group? Draw 5 more in each group, count how many there are altogether, then write the number in the box.

7

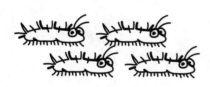

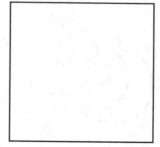

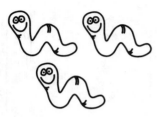

 **EXTRA!**

Ask students to say a number story for each row—"2 ants and 5 more makes 7."

Adding 5 More

How many pictures are in each group? Draw 5 more in each group, count how many there are altogether, then write the number in the box.

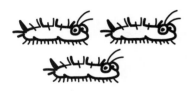

 **EXTRA!**

The last row on this page has no pictures. Help students to draw 5 bugs, count them, and write the number in the box.

NAME

Taking 1 Away

**Count the pictures in each group, then cross out 1 picture.
How many are left? Write the matching number on the line.**

 3  _____

 _____ 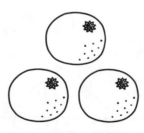 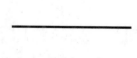 _____

 _____  _____

EXTRA! Help students to say a number story for each row—"4 apples take away 1 apple leaves 3 apples."

Taking 1 Away

Count the pictures in each group, then cross out 1 picture.
How many are left? Write the matching number on the line.

Are students familiar with "zero"? You may need to help them with the
last example on this page where they need to write "O" on the line.

Taking 2 Away

Count the pictures in each group, then cross out 2 pictures.
How many are left? Write the matching number on the line.

3 _____   _____

_____   _____

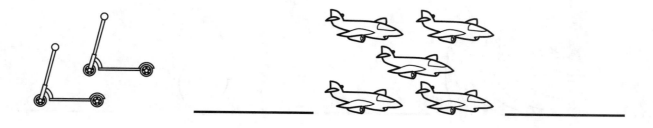

_____   _____

Help students to say a number story for each row—"5 cars take
away 2 cars leaves 3 cars."

Taking 2 Away

Count the pictures in each group, then cross out 2 pictures. How many are left? Write the matching number on the line.

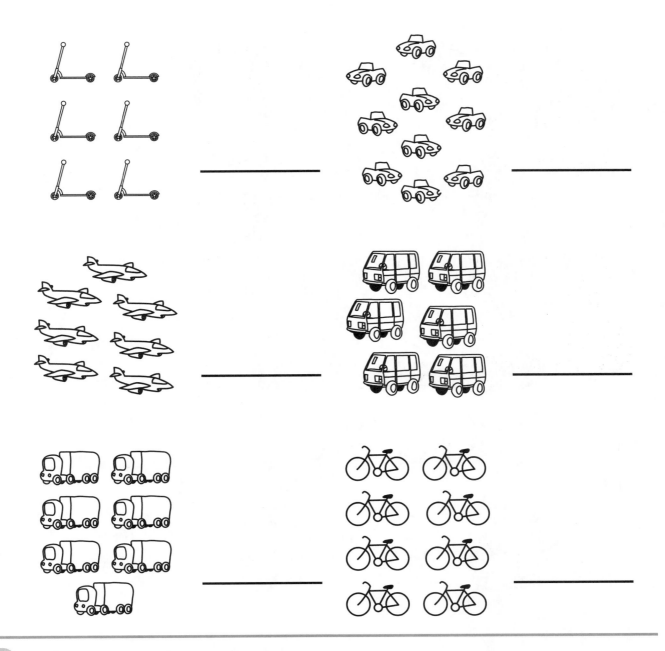

EXTRA! You can help students understand "taking away" by using real examples. For instance, if there are 4 bananas and you eat 2, how many are left?

#8105 Numbers to 10 • Adding • Subtracting

Taking 3 Away

Count the pictures in each group, then cross out 3 pictures. How many are left? Write the matching number in the box.

I

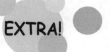

EXTRA!

Ask students to say a number story for each row—"4 lions take away 3 lions leaves 1 lion."

#8105 Numbers to 10 • Adding • Subtracting

Taking 3 Away

Count the pictures in each group, then cross out 3 pictures. How many are left? Write the matching number in the box.

 **EXTRA!** You can do this activity using dry pasta. Ask students to count them, take 3 away, and count them again.

#8105 Numbers to 10 • Adding • Subtracting

Taking 4 Away

Count the pictures in each group, then cross out 4 pictures. How many are left? Write the matching number in the box.

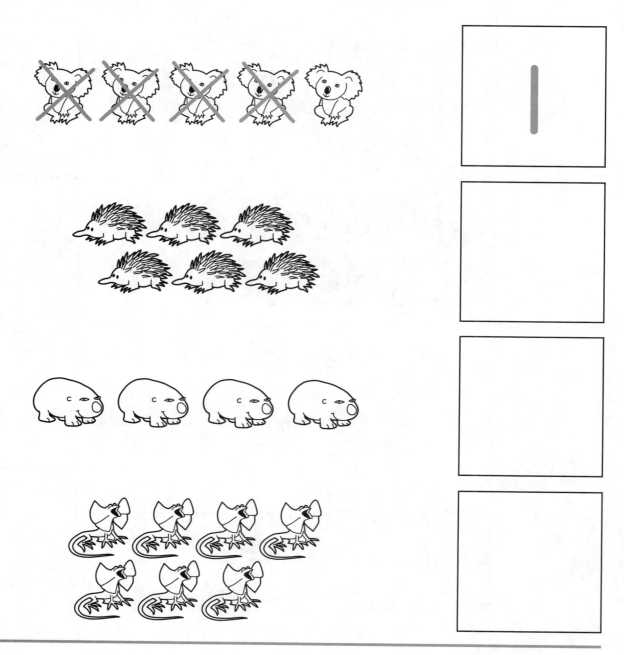

 **EXTRA!** Ask students to say a number story for each row—"5 koalas take away 4 koalas leaves 1 koala."

Taking 4 Away

**Count the pictures in each group, then cross out 4 pictures.
How many are left? Write the matching number in the box.**

 **EXTRA!** Do you know any simple counting songs or rhymes you can sing with students? See if you can find some that involve subtraction.

#8105 Numbers to 10 • Adding • Subtracting

NAME

Taking 5 Away

Count the pictures in each group, then cross out 5 pictures. How many are left? Write the matching number in the box.

2

 **EXTRA!**

Ask students to say a number story for each row—"7 horses take away 5 horses leaves 2 horses."

Taking 5 Away

Count the pictures in each group, then cross out 5 pictures.
How many are left? Write the matching number in the box.

---

(75)

#8105 Numbers to 10 • Adding • Subtracting

EXTRA!
You can repeat this activity with building blocks. Ask students to take a certain number away and count the remaining blocks.

Adding 1

How many fish are there in each group? Write this number in the first box. $\boxed{+}\,\boxed{1}$ means you have to draw 1 more fish, then count how many there are altogether. Write this number in the circle.

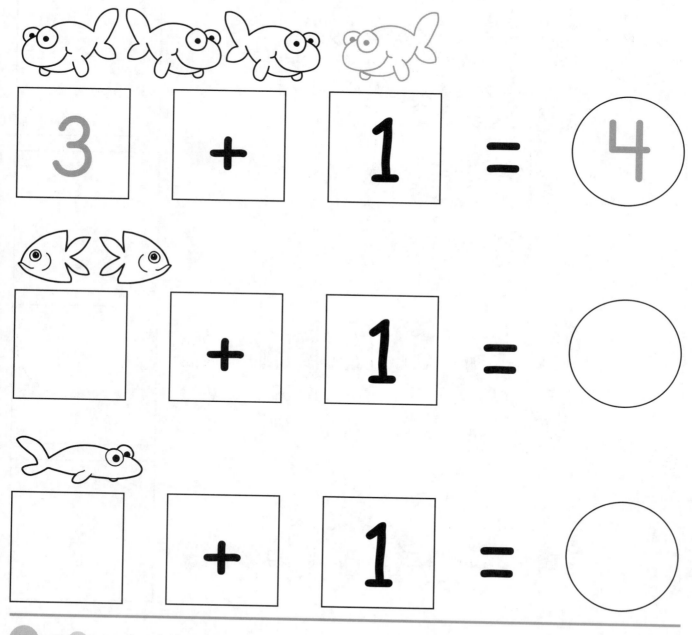

3 + 1 = 4

Ask students another addition question, for example, "If I have 2 fish and I add 1 fish, how many fish do I now have?" Students can count it using their fingers.

EXTRA!

#8105 Numbers to 10 • Adding • Subtracting

76

©Teacher Created Resources, Inc.

Taking 1 Away

How many fish are there in each group? Write this number in the first box. $\boxed{-}$ $\boxed{1}$ means you have to cross out 1 fish, then count how many fish are left. Write this number in the circle.

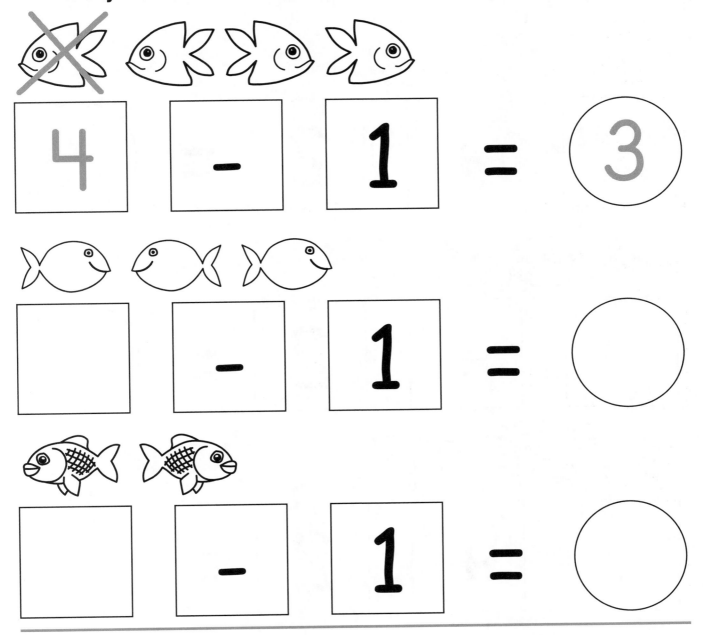

EXTRA! Try subtraction using a chalkboard or similar surface, e.g., draw 5 pictures, erase 1 and write how many are left.

©*Teacher Created Resources, Inc.*   *#8105 Numbers to 10 • Adding • Subtracting*

Adding 2

How many shells are there in each group? Write this number in the first box. [+] [2] means you have to draw 2 more shells, then count how many there are altogether. Write this number in the circle.

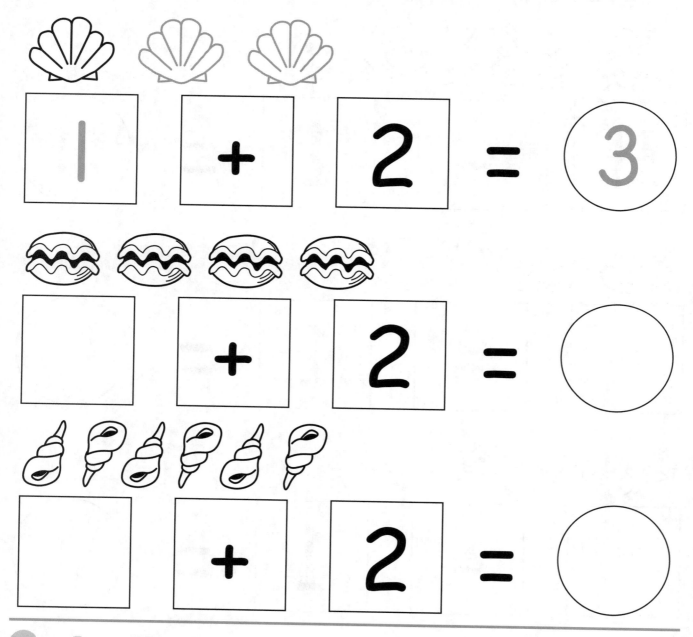

1	+	2	=	(3)

EXTRA!

Ask students another addition question, for example, "If I have 3 shells and I add 2 shells, how many shells do I now have?" Students can count using their fingers.

Taking 2 Away

How many shells are there in each group? Write this number in the first box. $-\boxed{2}$ means you have to cross out 2 shells, then count how many shells are left. Write this number in the circle.

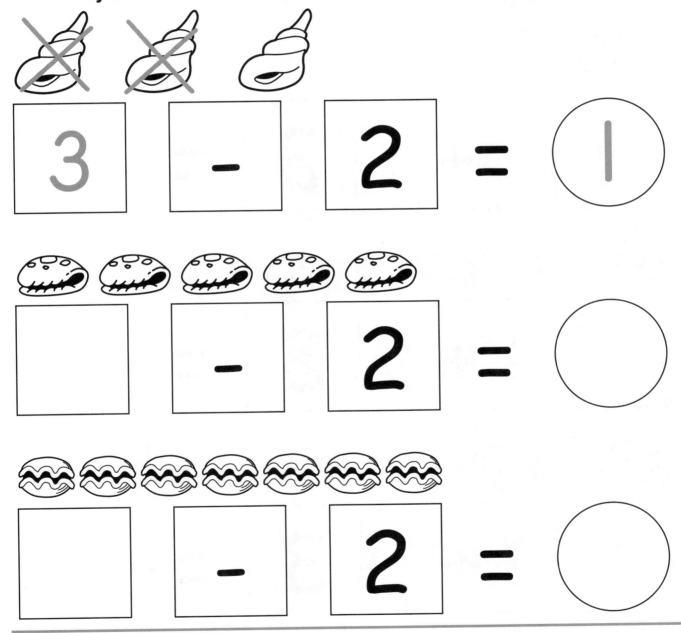

EXTRA!

Use a set of playing cards (using the cards from 2 to 10) to help students practice subtraction. Can they subtract one card from another and find the right number card for the answer?

#8105 Numbers to 10 • Adding • Subtracting

Adding 3

How many vegetables are there in each group? Write this number in the first box. $\boxed{+}\ \boxed{3}$ means you have to draw 3 more vegetables, then count how many there are altogether. Write this number in the circle.

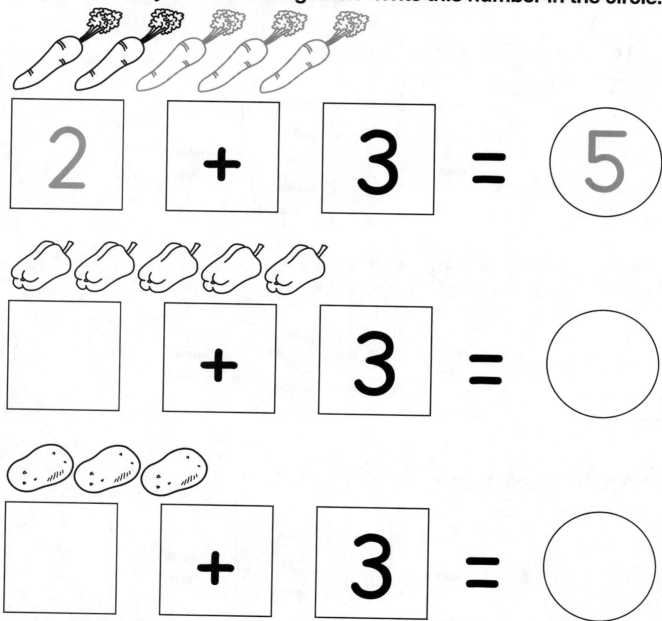

$2 + 3 = 5$

$\boxed{} + 3 = \bigcirc$

$\boxed{} + 3 = \bigcirc$

EXTRA! Ask students another addition question, for example, "If I have 5 peas and I add 3 peas, how many peas do I now have?" Students can count using their fingers.

Taking 3 Away

How many vegetables are there in each group? Write this number in the first box. $\boxed{-}\;\boxed{3}$ means you have to cross out 3 vegetables, then count how many vegetables are left. Write this number in the circle.

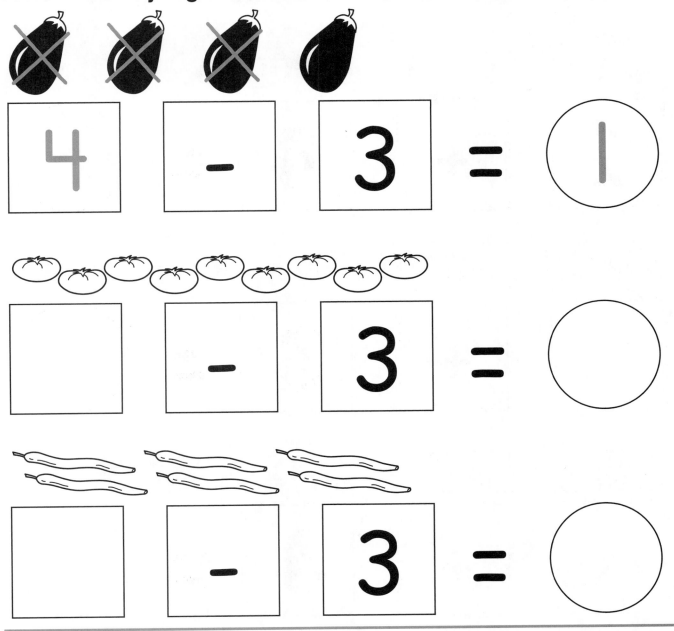

EXTRA!

Use a set of building blocks or small items and have students move them from one group to another and count them. If they take away 3 blocks, how many are left? If they add three, how many do they now have?

#8105 Numbers to 10 • Adding • Subtracting

Adding 4

How many pictures are there in each group? Write this number in the first box. $\boxed{+}$ $\boxed{4}$ means you have to draw 4 more pictures, then count how many there are altogether. Write this number in the circle.

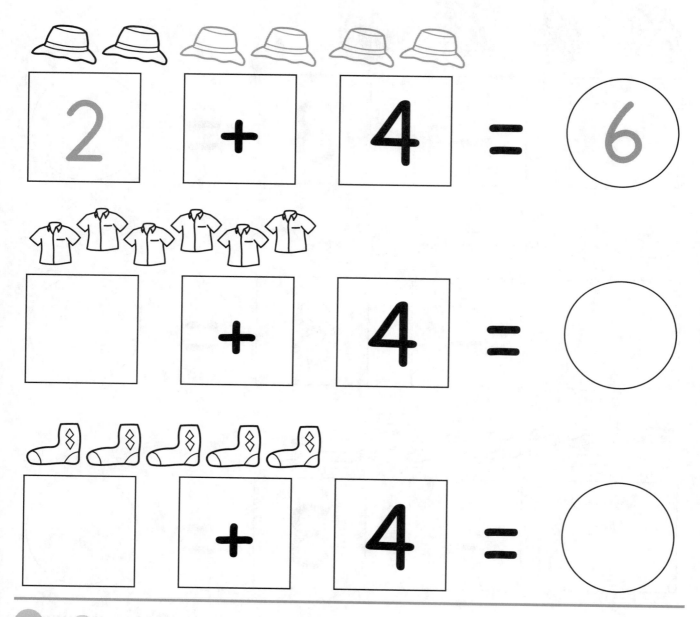

EXTRA!

Ask students another addition question, for example, "If I have 4 shirts and I buy 4 more shirts, how many shirts do I now have?" Students can count using their fingers.

Taking 4 Away

How many pictures are there in each group? Write this number in the first box. $\boxed{-}\;\boxed{4}$ means you have to cross out 4 pictures, then count how many pictures are left. Write this number in the circle.

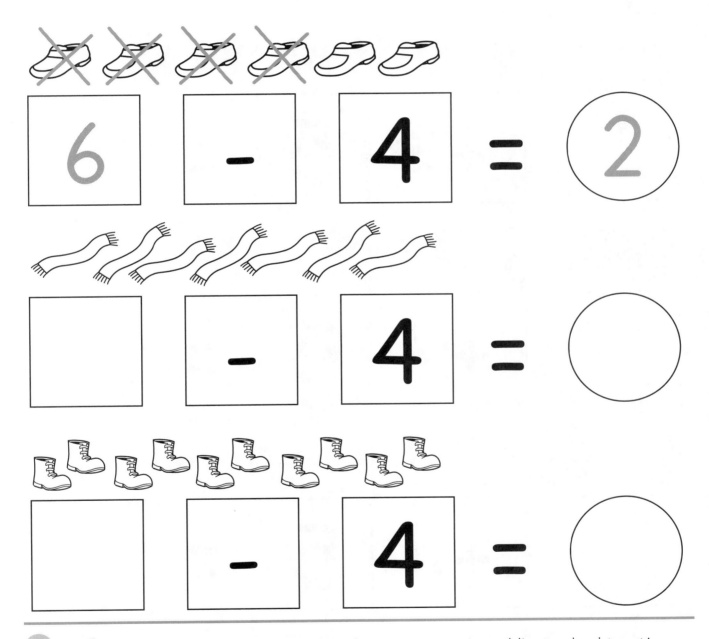

EXTRA!

Use groups of everyday objects to practice adding and subtracting. The more students can see how adding and subtracting are done, the easier it will be for them.

#8105 Numbers to 10 • Adding • Subtracting

Adding 5

How many aliens are there in each group? Write this number in the first box. $\boxed{+}\ \boxed{5}$ means you have to draw 5 more aliens, then count how many there are altogether. Write this number in the circle.

  Ask students another addition question, for example, "If I see 4 aliens and along come 5 more aliens, how many aliens are there now?" Students can count using their fingers.

Taking 5 Away

How many aliens are there in each group? Write this number in the first box. $\boxed{-}\boxed{5}$ means you have to cross out 5 aliens, then count how many aliens are left. Write this number in the circle.

Ask students another subtraction question, for example, "If I see 10 aliens and 5 go into their spaceship, how many aliens are left?" Students can count it using their fingers.

#8105 Numbers to 10 • Adding • Subtracting

Making Groups of 5

Each pizza should have 5 olives on it. Count the olives on each pizza. If you need more olives, draw them in to make 5, then write your number story in the boxes. If there are too many olives, cross some out to leave 5 and write your number story in the boxes.

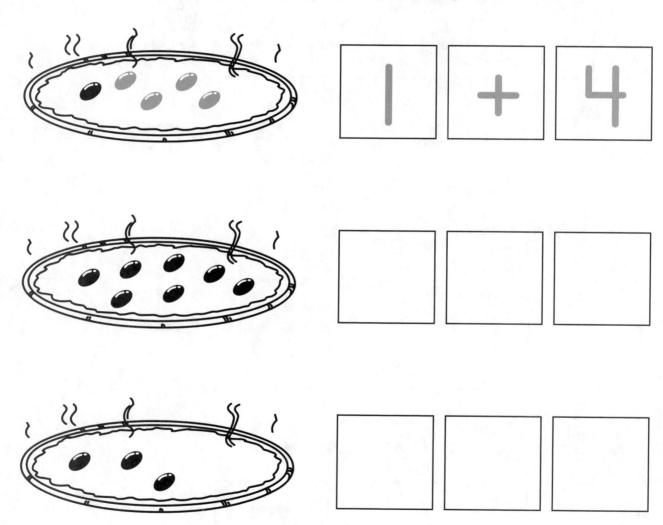

EXTRA!

Help students collect some counters (e.g., buttons). Put them into two groups. Can they count how many in each group, then count how many altogether?

Making Groups of 5

Each pizza should have 5 olives on it. Count the olives on each pizza. If you need more olives, draw them in to make 5, then write your number story in the boxes. If there are too many olives, cross some out to leave 5 and write your number story in the boxes.

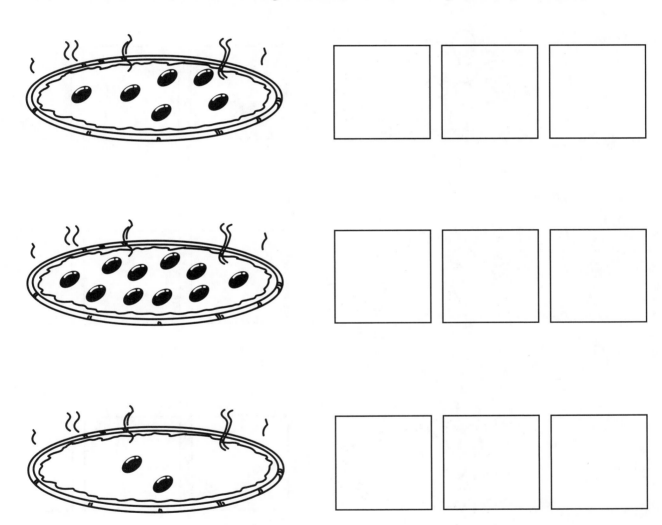

 **EXTRA!** Help students cut out pictures of food, then paste them into two groups. Discuss how many are in each group, and how many altogether.

NAME

Making Groups of 6

Each flower should have 6 petals on it. Count the petals on each flower. If you need more petals, draw them in to make 6, then write your number story in the boxes. If there are too many petals, cross some out to leave 6 and write your number story in the boxes.

7 − 1

EXTRA! How else can you make groups of 6? Find objects in the classroom and make groups of 6.

(88)

Making Groups of 6

Each flower should have 6 petals on it. Count the petals on each flower. If you need more petals, draw them in to make 6, then write your number story in the boxes. If there are too many petals, cross some out to leave 6 and write your number story in the boxes.

EXTRA! Take a handful of small objects and count them. Take some away and ask students to count how many are left.

89

Making Groups of 7

Here are some very spotty dogs! Each dog should have 7 spots on it. Count the spots on each dog. If you need more spots, draw them in to make 7, then write your number story in the boxes. If there are too many spots, cross some out to leave 7 and write your number story in the boxes.

 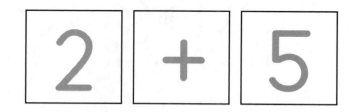

 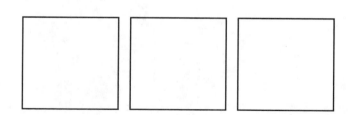

 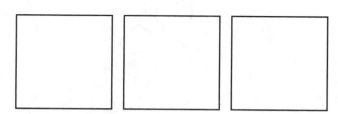

EXTRA! You could try this activity using a chalkboard and simple shapes. Students can draw more shapes or erase some to make a group of 7.

Making Groups of 7

Here are some very spotty dogs! Each dog should have 7 spots on it. Count the spots on each dog. If you need more spots, draw them in to make 7, then write your number story in the boxes. If there are too many spots, cross some out to leave 7 and write your number story in the boxes.

 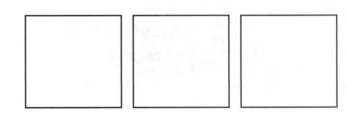

 Try to count the spots on a real cat or dog!

91

Making Groups of 8

How many legs does a spider have? Each spider should have 8 legs. Count the legs on each spider. If you need more legs, draw them in to make 8, then write your number story in the boxes. If there are too many legs, cross some out to leave 8 and write your number story in the boxes.

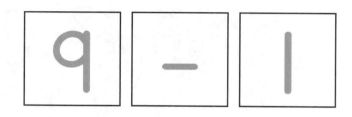

9	–	1

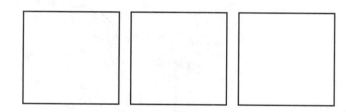

EXTRA!

Help students make spiders using colored paper. Use red and blue strips for legs. How many are red? How many are blue?

92

Making Groups of 8

How many legs does a spider have? Each spider should have 8 legs. Count the legs on each spider. If you need more legs, draw them in to make 8, then write your number story in the boxes. If there are too many legs, cross some out to leave 8 and write your number story in the boxes.

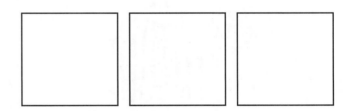

EXTRA! How many ways can you show 8 fingers? Help students find different ways (e.g., 5 fingers and 3 fingers).

#8105 Numbers to 10 • Adding • Subtracting

NAME

Making Groups of 9

Each tiger should have 9 stripes. Count the stripes on each tiger. If you need more stripes, draw them in to make 9, then write your number story in the boxes. If there are too many stripes, cross some out to leave 9 and write your number story in the boxes.

8	+	l

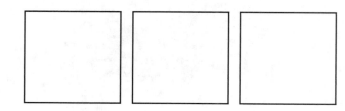

EXTRA!   Look for pictures or photos of tigers. How many stripes can students count on them?

Making Groups of 9

Each tiger should have 9 stripes. Count the stripes on each tiger. If you need more stripes, draw them in to make 9, then write your number story in the boxes. If there are too many stripes, cross some out to leave 9 and write your number story in the boxes.

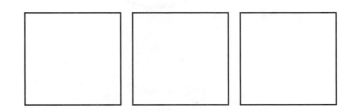

EXTRA! Draw the outline of a tiger, and help students paste on sticks for stripes. How many stripes will fit on the tiger?

(95)

Making Groups of 10

Each cat should have 10 whiskers. Count the whiskers on each cat. If you need more whiskers, draw them in to make 10, then write your number story in the boxes.

 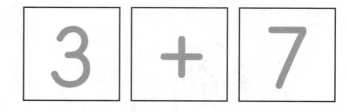

3	+	7

 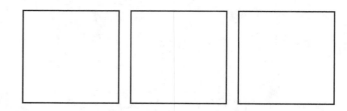

 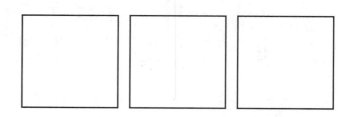

 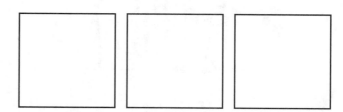

 **EXTRA!** Help students make a cat's face using a paper plate, then paste on sticks for whiskers.